Geysers and very hot springs in Europe:
6. Iceland

Geysers and very hot springs in Asia:
7. Kamchatka (Russia)
8. Tibet (China)
9. Tzu Peninsula (Japan)

6

7

8

9

12

11

10

Geysers and very hot springs in Australasia:

10. North Island (New Zealand)
11. New Britain and New Ireland
(Papua New Guinea)

Geysers and very hot springs in Africa:

12. East African Rift Valley

FACTS ABOUT GEYSERS

Geysers have so far been found in at least 40 places in the world. All of these are places where volcanos are active or have been active within the last few thousand years. The most active geyser field is at Yellowstone National Park, Wyoming, USA. It contains over 400 geysers including the world's most famous geyser – called Old Faithful, which shoots boiling water about every 70 minutes to a height of 150 ft – and the world's tallest geyser – Steamboat Geyser, which shoots water to over 380 ft. The largest amount of water gushes from Grand Geyser, Yellowstone, even though it does not produce a particularly tall plume.

The geyser which gave its name to all other geysers is Geysir, a geyser in Iceland that used to reach 180 ft. It is now just a quiet pool. Nearby Strokkur reaches 80 ft and is currently the tallest in Iceland. The largest geyser ever known was Waimangu, New Zealand. It holds the world record at 1500 ft – and is equivalent to the world's highest office building, the Sears Tower in Chicago, but it blew itself out, killing four people in 1917. The most spectacular areas for geysers after Yellowstone are now El Tatio in Chile and the Kronotski National Park, Kamchatka Peninsula, Russia.

The geyser fields in both Iceland and New Zealand have been much reduced as people have tapped into the boiling water supply to make steam for turning turbines to produce electricity.

 Grolier Educational Corporation
SHERMAN TURNPIKE, DANBURY, CONNECTICUT 06816

LAND O SHAPES
GEYSER

Author
Brian Knapp, BSc, PhD
Art Director
Duncan McCrae, BSc
Editor
Rita Owen
Illustrators
David Hardy and David Woodroffe
Print consultants
Landmark Production Consultants Ltd
Printed and bound in Hong Kong
Designed and produced by
EARTHSCAPE EDITIONS

First published in the USA in 1993 by
GROLIER EDUCATIONAL CORPORATION,
Sherman Turnpike, Danbury, CT 06816

Library of Congress #92–072045

Cataloging information may be obtained
directly from Grolier Educational Corporation

Title ISBN   0–7172–7183–8

Set ISBN   0–7172–7176–5

Acknowledgements. The publishers would
like to thank the following: Redlands County
Primary School.

Picture credits. All photographs from the
Earthscape Editions photographic library except the
following: NASA 14t; ZEFA 25t, 35tr; Hutchison 35b

Cover picture: Castle Geyser, Yellowstone
National Park, Wyoming, USA.
Inside back cover picture: Morning Glory Pool,
Yellowstone National Park, Wyoming, USA.
Page 5: Old Faithful, Yellowstone National Park,
Wyoming, USA.

In this book you will find some words that have been shown in **bold** type. There is a full explanation of each of these words on page 36.

On many pages you will find experiments that you might like to try for yourself. They have been put in a blue box like this.

In this book mi means miles and ft means feet.

These people appear on a number of pages to help you to know the size of some landshapes.

CONTENTS

Take care with Geysers

Geysers and hot springs are some of the world's most interesting landshapes and you are sure to want to visit them. However, most geysers and hot springs contain water that is hot enough to scald you. It is not safe to stand close to a geyser or even to put a finger into the water of a hot spring to test its temperature.

Introduction

Geysers are springs that send powerful jets of steam and water into the air from time to time. They are found in places where rocks quite close to the surface have been heated. Many geysers therefore are found in the same areas as volcanos or where volcanos were once active.

Geysers are not the only signs that there are hot rocks near the surface. Bubbling pools, hot springs and mud pools are in fact much more common than geysers. Sometimes when there is little water available puffs of steam or gases with breathtaking smells are the only signs. These are called **fumaroles**.

Hot rocks are also places where great chemical changes occur. As steam, gases and water move through the rock they dissolve many of the **minerals** from the rock, so the gushing fountains and bubbling pools are far from being pure water. Rather, they contain a wealth of dissolved materials.

As the hot, mineral-laden water comes to the surface it cools, depositing the minerals on the land around. This is where the landshaping begins, because the minerals build up into some of the world's most beautiful and fantastic forms.

In this book you can find out about the many landshapes produced by geysers and hot water, and you can enjoy the many shapes and patterns they produce. Simply turn to a page of your choice and enjoy the world of geysers.

Chapter 1:
How geysers work

Where geysers are found

Geysers are usually found with **spouters**, hot springs, mud pots and fumaroles, making a region called a geyser field.

Each feature of the geyser field is produced as a result of water travelling between hot underground rocks. Which type of feature is found in any geyser field simply depends on the local underground 'plumbing'.

Geyser fields are usually easy to spot because the ground surrounding them is often clear of all plants and the surface of the ground is a greyish-white color.

Hot springs and pools occur when the 'plumbing' allows water to flow out easily (see pages 22 to 24).

Variety in a geyser field

A geyser field usually has steam curling skywards from many places. This shows where the underground system is 'leaking' heat. Some of the leaks form bubbling springs or pools, others produce geysers, and yet more will just continue to puff steam because they do not have a good enough water supply.

In some places water seeps out as springs and terraced pools are made of the mineral deposits (see page 26).

The scalding hot water is first heated deep underground (see pages 14 to 16).

Geysers often build up nozzles (see page 18).

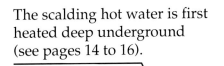

Geysers erupt intermittently (see page 12).

Mud pots (see page 30) and steamers (see page 28) are found when there is no continuous flow of water from underground.

The greyish-white deposits are from rocks dissolved deep underground (see page 20).

How a geyser erupts

Geysers regularly spurt water high into the air.

You know *where* a geyser will erupt because there is a hole in the ground or a raised nozzle through which it will spurt. However, because the geyser's 'plumbing' is hidden deep below the surface, it is very difficult to tell *when* a geyser will erupt. This makes geyser-watching even more exciting.

The first signs

When a geyser erupts it begins by sending small spurts of water into the air, rather like a pan of water that is about to boil over. Gradually the spurts of boiling water get more powerful and rise higher and higher, but always there is a pause between each one.

Getting stronger

No one knows just how high the geyser will reach each time, but some of the world's biggest geysers regularly reach over 150 ft in height.

As the pulses of water get stronger and the geyser gets close to its main eruption, the gap between the pulses gets shorter. As the water fountain gets higher it gets caught by the wind and a spray starts to float in the air.

The main event

The geyser will burst into full action almost without warning. Huge plumes of water rise into the air and there is a great roaring sound like an immensely powerful fire hose. It is quite a spectacular sight.

The cycle ends

Geysers can keep their fountains playing for many minutes, continually sending spurts of water that rise and fall in the air. Then they literally run out of steam and quickly fade away.

Heating the water

Geyser fields depend on a source of hot rocks deep underground. Most of these rocks are found close to places where volcanos have erupted, because the chamber of molten rock that fed a volcano can remain hot for thousands of years after an eruption.

When volcanos erupt they often shake the ground violently, causing the rocks to crack. Later, when the eruption has died away, rainwater can seep down into the cracked rock and become heated.

This picture shows a typical site for the location of hot rocks. It is part of the Rocky Mountains surrounding the Yellowstone geyser fields in Wyoming. It has been taken from high up and the view is almost like a map.

Yellowstone has been ringed to show its position more clearly. There used to be a huge volcano at this spot, but long ago it exploded so violently that the volcano collapsed. All the geyser fields are sited on the top of the old volcano because there is still a large area of hot rocks below. Indeed, it is even possible that the volcano will erupt again one day.

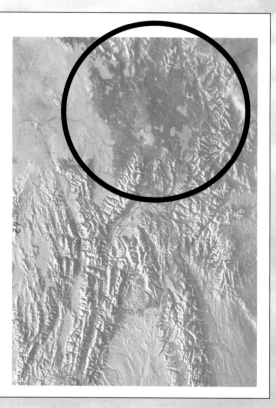

Water needs
Geysers use enormous amounts of hot water and therefore occur in areas where there is either a high rainfall or snowfall, such as in a mountain region.

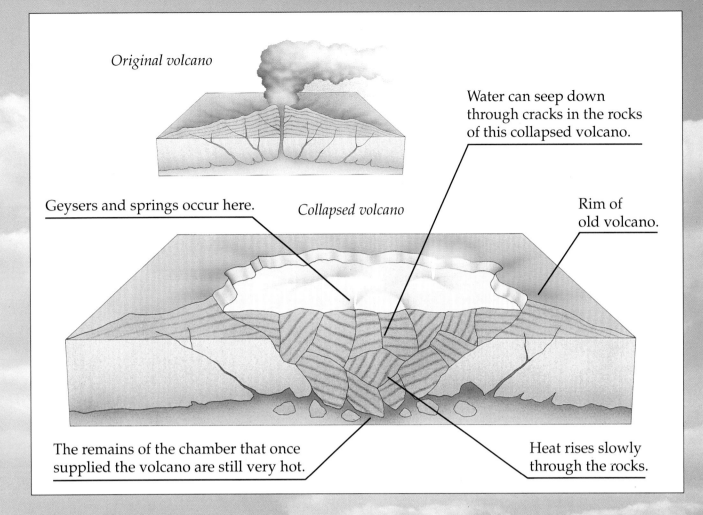

Original volcano

Water can seep down through cracks in the rocks of this collapsed volcano.

Geysers and springs occur here.

Collapsed volcano

Rim of old volcano.

The remains of the chamber that once supplied the volcano are still very hot.

Heat rises slowly through the rocks.

The best rocks

Geysers can only form where water moves easily underground. The rocks have to have many gaps or cracks, called **fissures**, which let rainwater or melting snow trickle and seep into the ground.

   The trickling and seeping takes a long time – perhaps several hundred years. But there also have to be much bigger cracks to carry the heated water up to the surface in just a few minutes.

For more information about volcanos and magma chambers see the book Volcano *in the Landshapes set.*

Geyser 'plumbing'

Geysers need to have a very special underground 'plumbing' system in order to spurt water and steam high into the air. They need a source of heat, an underground chamber, or reservoir, where water can gather and be heated, and a wide 'pipe' to get the heated water quickly to the surface. But the key to the whole system is that there has to be a narrow section in the 'pipe' near the ground's surface. It is only by chance that all these features are found occurring together, which explains why geysers are rare.

Make a geyser

This activity must only be done by an adult, but it will be fun for you to watch. The adult should get a 3 ft long plastic tube and push a clear funnel on to one end. They should push the other end through a cork with a hole in it (like the ones used in wine making).

Warning.

You must be very careful because the water will boil. An adult must be present at all times and take responsibility for the safety of the experiment. Use protective gloves, clothes and goggles to prevent damage from steam and hot water splashes.

The cork should be fitted into the top of a boiling flask and then the funnel supported about 30 inches above the bottom of the flask. On no account hold the funnel in the hand. The tube, flask and funnel are then filled with cold water until the funnel is half full.

The flask is then heated using a gas flame such as a camping stove or bunsen burner. Stand well clear.

Eventually the water in the flask will boil. The size of the eruption will depend on how the water is heated and the height of the funnel above the flask. The higher the tube is raised, the more spectacular the eruption.

Keeping the lid on

All geysers must have a narrow section in their plumbing, usually close to the surface. This narrow section stops the **superheated water** from escaping until enough pressure has been built up. If there is no narrow part then the water flows out continuously as a hot spring. If the narrow section is not quite small enough the water forms a 'spouter'.

HOW THE ERUPTION HAPPENS

FEATURES OF THE 'PLUMBING'

6. The steam and water shoot out of the geyser until there is no steam left.

Water and steam.

ground level

5. Eventually the water pressure gets so high it begins to boil. This causes cold water to be thrown out as a fountain along with the hot water which has turned to steam.

Narrow passageway.

Narrow passageways can measure from a few inches in the smallest geysers to two feet in the largest.

4. The weight of cold water seeping into the upper passageway from above acts like a lid as on a pressure cooker. The water below gets hotter and hotter but it cannot escape.

Chamber

The passageways become lined with a scale made of dissolved minerals keeping them water-tight and narrow.

3. Heated passageways continue to heat the trapped water.

Narrow passageway.

For water to move in and out of the ground easily the rock must be heavily cracked and the cracks must be large. If the cracks are small the water will move too slowly and most of the heat will be lost before it reaches the surface.

2. Water heats up above boiling point in a chamber that has been dissolved away.

Chamber

1. Water seeps through the rocks and builds up in underground passageways.

2000 – 3000 metres from surface to lowest collecting chamber.

This diagram is not drawn to scale.

The rocks are hot at this great depth.

17

Nozzles and platforms

Water from a geyser is guided out of the ground by a **nozzle** in much the same way as water comes out of a fire hose; the smaller the nozzle, the greater the pressure builds up in the passageways and the higher the geyser erupts.

The nozzle is formed from the dissolved minerals as they are deposited at the surface. Sometimes it only resembles a low mound, at other times it looks more like the turret of a castle. As a result, many geysers are named after the shape of their nozzles. The nozzle rises from a broader mound of grey rock called a **platform**.

This is the nozzle.

Castle Geyser (above) has a nozzle that reminds people of a castle tower. As the nozzle is wide the water escapes from it quite easily.

The water from the geyser is rich in minerals. These provide nutrients for tiny plants called algae. Algae give to the rocks on which they are growing their brown, yellow and green colors.

This is the platform for Grand Geyser.

Cones and fountains

Geysers gush forth in many different ways because of their internal plumbing: the bigger the chambers in the plumbing system, the longer the geyser erupts.

The shape and size of the water plume depends on the shape of the hole through which the water finally gushes in just the same way as the pattern of spray can be changed by altering the nozzle at the end of a garden hose.

For geysers the nozzle can either be a **cone**- shape (which gives a tall plume eruption) or a **crater**-shape (which gives a broad fountain effect). Most cones grow by up to an inch a century.

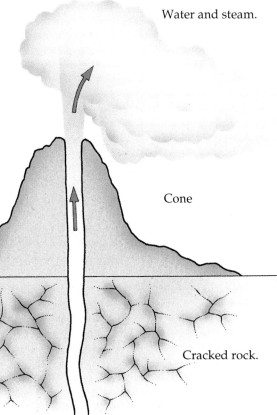

Water and steam.

Cone

Cracked rock.

Cones

Cone-type geysers have a cone-shaped nozzle at the surface and a narrow section of 'plumbing' below the surface. Old Faithful (see page 32) and Castle Geyser (see page 18) are examples of cone-type geysers.

Holey cones

The sudden growth of a geyser in a forest will kill all the trees near the nozzle. Then, as the nozzle grows, it will deposit its material around the tree stumps.

When the trees finally rot away they leave holes in the geyser nozzle like the ones shown here.

How cones form

The water gushing from a geyser cools very quickly once it reaches the surface. As it does so the dissolved materials are deposited in a similar way to the lime scale that is deposited on the sides of a kettle.

The geyser scale forms around the cracks that bring water to the surface. The scale gradually builds up layers that eventually form the **sinter** cone of the geyser.

When the nozzle grows too high, the force of the spurting water may break pieces off the edges which fall away. As a result, nozzles rarely grow to more than a few yards in height.

The largest cone in Yellowstone National Park is Castle Geyser (page 19). The cone is more than 5000 years old.

A piece of sinter from a geyser. You can see the cast of the wooden twigs that have long since rotted away.

Fountains

Instead of a nozzle, fountain type geysers have a crater at the surface. As a result there is nothing to direct the water jet, and there is just a great outpouring of spray. This type of geyser is much more common than the cone type. Fountain types build broad sinter platforms as the dissolved minerals are deposited over a wide area.

21

Chapter 2:
Hot springs

Bubbling pools

These form when hot water can reach the surface as a regular flow. They are really hot springs. The water still moves upwards as it superheats and expands, but it is not trapped below a column of cold water.

As a result the water reaches the surface quietly, steadily dissolving away underground rocks and depositing the dissolved materials as a rim around the pool. In this way the pool grows deeper.

Most pools have clear water that is close to the boiling point. This is the Morning Glory Pool, Yellowstone. You can see a bigger version of this beautiful pool on the inside rear covers of this book.

Beyond the rim the water cools in the air and algae will grow on the deposits. They make the beautiful orange colors.

This picture shows a close-up view of a rim.

How a rim grows
As the water comes to the surface it cools and deposits many of the dissolved minerals forming a crust on the surrounding rocks.

Gradually the crust may build into a rim and the spring becomes trapped into a pool.

The pools are crystal clear because the water is too hot for most life forms.

The world's biggest hot pool

Grand Prismatic Hot Spring in Yellowstone National Park is the largest hot spring in the world. It shows clearly how rising water filled with minerals can build up huge low-angled domes.

The pool surface, seen here as a brilliant blue color, steams throughout the year, while the water seeps over the edge and towards the bank shown in the picture below.

Rocky 'waterfall'

As the water spills out of the pool it begins to form a wide river with many channels. All the time the water is cooling and it can hold less and less of the dissolved minerals it brought to the surface.

As the minerals are deposited in the channel, they form layers that gradually build up one on top of the next. Here you can see it spilling down a bank and looking just like a waterfall.

The view from the air shows just how wide the pool is. The pool is very shallow and its rim has broken away in many places so that water spills in every direction. As more and more mineral is deposited, so the dome gets bigger and bigger.

Glistening flowstone

Most hot springs flow up through rocks that are slow to dissolve. But where hot water seeps through limestone, the rock dissolves quite quickly and the surface near the springs becomes covered in vast deposits of white limestone. On steep slopes they build into the most amazing white benches or terraces.

How flowstone pools form
The terraces build rims where the water most easily evaporates. Once a rim has begun to form, the water tends to pond up behind, making a shallow pool.

Flowstone

As the water drips over the edges of the pools and evaporates it deposits limestone. Over the years the limestone builds spectacular features such as long columns, or hanging cones similar to the features found in a limestone cave.

All these features are called flowstone because they formed due to flowing water.

What causes the sparkle and color?

Most of the mineral seen here is a form of limestone, which is white. It sparkles in the sunlight because the limestone is deposited as tiny crystals. The light bounces, or reflects, off the crystal surfaces.

The colors are produced by traces of other minerals that were dissolved by the hot water as it seeped through the ground.

Sulphur and iron are the most common staining minerals. The sulphur stains the hillside with bright yellow streaks, iron stains with oranges, reds and browns.

Chapter 3:
Steamers and mud pots

Steamers

A steamer, or fumarole, is the name given to a kind of hot spring where there is very little water. Some people refer to fumaroles as steam vents. Because there is so little water, as soon as it touches the hot rocks, all of it is simply turned into steam.

Fumaroles are the hottest of all the springs because the steam reaches the surface rapidly and because small amounts of water are quickly heated.

A region with little water may have mainly steam fumaroles. This is the Bumpus Hell Basin on Mt Lassen, California, USA. In the picture below mud pools can be seen in the lowest areas and steam vents farther up the slopes.

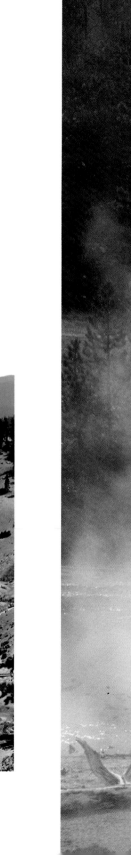

Smells of 'rotten eggs'

A number of gases, such as odorless and colorless carbon dioxide are combined with steam coming from geysers, hot springs, mud pots and steam vents. However, when the amount of water is small the rotten eggs smell is most noticeable. It is caused by a gas called hydrogen sulphide.

This is a view down the throat of a steamer. Notice how the steam has **corroded** the rock, making the throat bigger.

Mud pots

Mud pots form where rainwater or snowmelt fill in the vent of a steamer. The gases from the steamer become dissolved in the water and make the water very acid. This allows the mud to attack the surrounding rock, enlarging the surface area into a deeper and deeper pool or pot. The mud in the pot is actually a mixture of the spring water and corroded rock.

You may be surprised to know that bacteria can live in this 'devil's cauldron' despite its acid water.

The bubbles and spatters you see occur every time the gases in the steamer build up enough pressure to burst their way through the mud to the surface.

Mudpools, unlike geysers and hot springs, change through the year depending on how much rainfall or snowmelt there has been. For example, in spring just after snowmelt they are much more runny than in the parched conditions of late summer.

As hydrogen sulphide bubbles into the water of a mud pot it makes sulphuric acid – the same as the acid in a car battery. Heat and acid corrode the rock walls very quickly.

Chapter 4:
Geysers of the world

Yellowstone

The land of a thousand smokes is the name given to the area of Yellowstone National Park in Wyoming, USA, by the Indians who first lived in the area.

This is the largest US National Park, covering 3468 square miles, but the geysers, hot springs and terraces are concentrated in several small areas. There are about 200 geysers and 10,000 hot springs and numerous fumaroles.

Old Faithful has erupted over a million times since it was first seen by explorers in 1870.

The Yellowstone River makes its way between the steamers and mudpools near Old Faithful. Its waters are rich in minerals and stay warm throughout the year, an ideal place for plants and small fish.

Famous and spectacular
Yellowstone has the world's most reliable geyser, called Old Faithful. This geyser sends a plume of water and steam up to 150 ft into the air.

It does not, however, erupt quite as regularly as its name suggests and the spacing between eruptions depends on how much water is ejected. A powerful eruption will completely empty the underground water chamber and because it will take longer to fill, the following eruption will be delayed. The time between eruptions is normally 50 to 80 minutes.

Other geyser fields

Because such special conditions are needed before geysers can form, important geyser fields are rare. Some of the best known are in New Zealand and in Iceland.

Geysers are areas of natural energy, and people have often used this power to make electricity. But you cannot have spectacular geyser fields and produce electricity. Thankfully, Yellowstone remains unused and therefore is as spectacular as ever. In New Zealand and Iceland, however, the geysers are no longer as spectacular as they once were because their waters have been tapped for electricity use.

Iceland

The word geyser comes from the Old Norse word goysa which in modern Icelandic is *geysir*, which means to gush. Iceland is part of a huge undersea volcano. It is also a place that receives plentiful rain or snow. Most of the island has hot springs, and all the homes in Reykjavik, the capital of Iceland are heated by natural hot water.

Power is obtained by harnessing the underground heat. In the photograph on this page you can see a power plant for harnessing the heat.

Iceland is the home of the famous Great Geysir, which used to erupt to a height of 230 ft every six hours. Unfortunately Great Geysir no longer erupts and the biggest geyser on Iceland is Strokkur (the Churner) which rises to 80 ft every quarter of an hour.

New Zealand

In the North Island of New Zealand there is a region that still has active volcanos. In the Whakarwarewara Thermal Reserve, Rotorua the largest geyser ever known to erupt, called Waimangu, used to erupt to 1500 ft every 6 to 15 hours, but it blew itself out in 1917.

Today there are several major geysers, of which the best known are Pohutu, a fountain type geyser, and Wairoa, a cone type geyser, seen erupting in the picture below.

New words

cone

the dome-shaped build-up of sinter that forms at the place where some geysers erupt

corrode

the acid action of gases and hot water that can dissolve rocks and wear away passages for the water to move through. Most geysers, hot springs, mud pots and fumaroles make their passages wider with this hot acid

crater

the pit-shaped depression that forms at the place where many fountain geysers erupt

fissure

any large split in a rock. Fissures can widen to make gaps, or they can remain closed. In a geyser, fissures allow heated water to get through the rocks and corrode them

flowstone

the name given to the deposits of limestone from hot springs and pools which builds up in smooth flowing shapes to cover the surface rocks below some hot springs. It can look like a frozen waterfall

fumarole

the name for the rising steam that comes from some openings in a geyser field. Fumaroles only spout steam because there is not enough water for pools or geysers

mineral

a natural material from a rock that can form crystals. The minerals dissolved in the water that comes from geysers and hot springs builds up deposits. Tiny crystals can be seen in these deposits

nozzle

the name for the build-up of minerals around the place where the geyser spurts from the ground. Geyser shapes affect the way the geyser erupts

platform

the broad, gently-sloping surface that surrounds a geyser. It is built from minerals that are deposited as the geyser water cools when it flows away from the nozzle

sinter

the name given to the quartz scale that makes nozzles and platforms. It is made from the mineral quartz, the same mineral found in some types of sand

spouter

a geyser which continuously boils over without ever really making a tall plume of water. It is half way between being a pool and a true geyser

superheated water

water above its normal boiling temperature. The water seeping through the rocks 6000 ft below the surface is under so much pressure that it may get as hot as 400 °F and still be a liquid. (At the surface water boils and becomes steam at 212 °F). When superheated water reaches the surface it almost explodes into steam. This is what gives the geyser its power to jet high into the air

Index

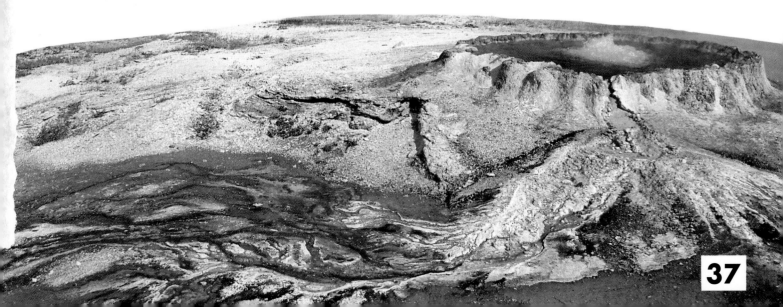

Morning Glory Pool, Yellowstone National Park, Wyoming, USA. This beautifully-colored pool is one of the main attractions of the National Park.